GEOLOGY FOR KIDS

GEOLOGY FOR KIDS

A JUNIOR SCIENTIST'S GUIDE
to Rocks, Minerals, and the Earth Beneath Our Feet

JUNIOR SCIENTISTS

MEGHAN VESTAL

callisto publishing
an imprint of Sourcebooks

Thank you to my husband, Ben, who always supports me and my big dreams.

Callisto Kids,
An imprint of Callisto Publishing LLC

Published by Callisto Publishing LLC C/O Sourcebooks LLC
1935 Brookdale Rd, Suite 139, Naperville, IL 60563
(630) 961-3900
callistopublishing.com

Text by Meghan Vestal
Photography: Tom Pfeiffer, cover; Daniel Prudek/iStock, pp. ii, 27; Scott Camazine/Alamy, pp. vi, 70; RYSZARD FILIPOWICZ/iStock, pp. viii, 1; andyKRAKOVSKI/iStock, p 7; terrababy/iStock, p 15; Mlenny/iStock, pp. 16, 17; sara_winter/iStock, p 18; hadynyah/iStock, p 19; YayaErnst/iStock, p 20; Chen-Zheng-Feng/iStock, p 20; photosbyjim/iStock, p 21; den-belitsky/iStock, p 22; chaolik/iStock, p 22; ARCTIC IMAGES/Alamy, p 23; SumikoPhoto/iStock, p 23; Riaan Albrecht/Alamy, p 24; JurgaR/iStock, p 24; Sulo Letta/Alamy, p 25; DEDY DAMIYANTO/iStock, pp. 28, 29; C. Storz/Alamy, p 30; kellyvandellen/iStock, p 31; All Canada Photos/alamy, p 31; EyeEm/Alamy, p 32; PF-(usna)/Alamy, p 34; Zacarias Pereira da Mata/Alamy, p 35; Mint Images Limited/Alamy, p 37; pum_eva/iStock, p 41; Matt Kilroy/iStock, p 42; benedek/iStock, p 45; Jon Davison/Alamy, pp. 46, 47; THE NATURAL HISTORY MUSEUM/Science Source, p 49; olpo/Shutterstock, pp. 49, 74; domenico farone/Alamy, pp. 49, 66, 75; Shawn Hempel/Alamy, p 49; Myriam B/Shutterstock, pp. 49, 66, 75; Susan E. Degginger/Alamy, pp. 51, 52, 53, 54, 69, 70, 71, 72, 73; Elena Batkova/iStock, pp. 51, 69, 70; mahirart/Shutterstock, p 51; Aleksandr Kalugin/Alamy, p 51; bjdlzx/iStock, pp. 52, 73; BRUCE BECK/alamy, pp. 54, 71; E.R. DEGGINGER/Science Source, pp. 54, 72; Dominique Braud/Dembinsky Photo Associates/Alamy, p 55; Roberto Nistri/Alamy, p 55; jatrax/iStock, p 59; Gudella/iStock, pp. 60, 61; David Chapman/Alamy, pp. 65, 74; TOM McHUGH/Science Source, p 66; Nuttanin Kanakornboonyawat/Alamy, p 66; geogphotos/Alamy, p 66; SPL/Science Source, p 66; Buiten-Beeld/alamy, p 68; mrpluck/iStock, p 68; Gerry Bishop/Alamy, p 71; Roland Bouvier/Alamy, p 72; Bob Gibbons/Alamy, p 74; blickwinkel/Alamy, p 75; Obradovic/iStock, p 77; Bo Jansson/Alamy, p 78.

Interior and Cover Designer: Regina Stadnik
Photo Art Director/Art Manager: Tom Hood
Editor: Mary Colgan
Production Editor: Ashley Polikoff

The authorized representative in the EU for product safety and compliance is Dorling Kindersley, Verlag GmbH., Arnulfstr. 124, 80636 Munich, Germany.

Source of Production: Wing King Tong Paper Products Co. Ltd. Shenzhen, Guangdong Province, China
Date of Production: April 2026
Run Number: 5065469

Printed and bound in China
WKT 2

CONTENTS

Basalt sea cave on
Akun Island in Alaska

WELCOME, JUNIOR SCIENTIST!

Have you ever wondered what the inside of Earth looks like or where mountains came from? Are you curious about what kinds of rocks are in your backyard? Do you dream of one day finding a fossil? Where could you find one? If you ask questions like these and want to know the answers, then you are a junior scientist!

This book will explore Earth and its deepest, darkest (and hottest) secrets! Every amazing natural feature you see around you is the result of millions of years of change. These changes are still happening today—just very, very slowly. Mountains are getting higher, valleys are getting deeper, and rocks are being created and destroyed. This book will explain how. You will also get your hands dirty with simple experiments and activities you can do at home.

If you have a blank notebook or pad, keep it handy. You can use it as a geology journal to record things you find interesting and to keep track of any rock samples you collect. Ready to learn about what makes Earth rock and roll? Put on your geologist hat and let's dig in!

Ama Dablam mountain in the eastern Himalayan range, Nepal

Chapter One

EARTH, HOME SWEET HOME

Where is home? People, animals, and all the living things that we know of call Earth home. Scientists have been studying our planet for centuries and they learn new things every day. The study of Earth, what it is made of, and its history is called **geology**. Scientists in the field of geology are called geologists. As a junior scientist, you can be a geologist, too!

What Is Geology?

If you've ever looked at rocks or dug in the dirt, you've been a junior geologist. Geologists study:

- what Earth is made of
- Earth's history
- Earth's features, or landforms, like mountains, oceans, and deserts
- natural hazards, like earthquakes and landslides
- how Earth has changed over time
- Earth's future

They also learn things about what is happening deep inside Earth. Geologists learn about these things by studying rocks, fossils, and dirt. All the information they find helps them predict and prepare for things like earthquakes and volcanic eruptions that may happen in the future. It also helps them learn ways to protect our planet and its environment.

The Birth of the Earth

Have you ever noticed specks of dust in the air? Can you see any now? You may be surprised to learn that our entire planet began as tiny specks, or particles, in space.

Scientists believe that Earth was formed about 4.6 billion years ago. A force called gravity caused gas and dust in space to pull together. As the particles moved, they crashed into each other and clumped together. They began to form a hot ball of rock. This ball of rock became the planet you live on!

The ball of rock grew as more particles and other rocks were sucked in. The new planet—Earth—grew bigger and bigger, and hotter and hotter. The densest parts sank to the middle and became the **core**. The rest of the material became two layers—the **mantle** and the **crust**. We'll talk about Earth's layers more in chapter 2.

An Ever-Changing Planet

Earth has been changing ever since. Lakes, rivers, and oceans shrink and grow. Mountains rise and fall. Land moves and takes new forms. These changes are caused by many different forces. Some are forces that we can see and feel, like wind, rain, ice, and earthquakes. Others are forces that we cannot see because they happen far beneath the surface.

WEATHERING AND EROSION

Two processes that change the planet's surface are weathering and erosion. **Weathering** is when rocks and minerals are broken into smaller parts. This can be caused by natural forces

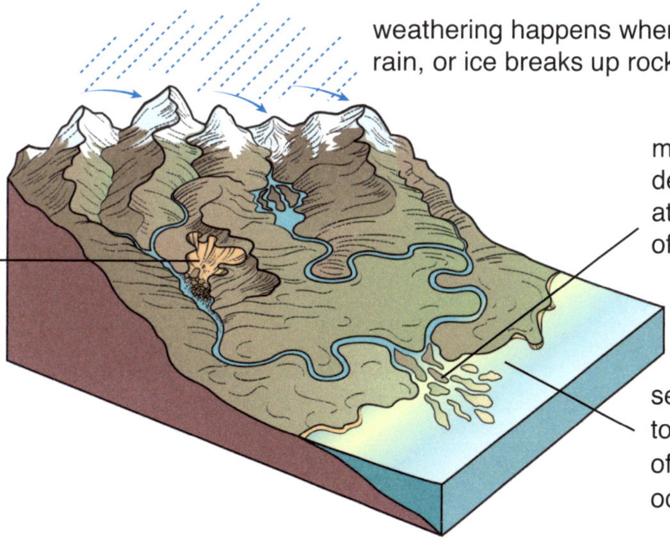

like wind, water, and ice. The small pieces created by weathering are called **sediments**. Here are some examples of weathering:

- Water in a river smooths rocks and soil as it flows over them.
- Wind blows pieces of rock off a cliff or mountain.
- Water gets into the cracks of a rock and freezes. It expands, or gets bigger, as it freezes and breaks the rock.

Do you remember the last time it rained? Did you notice what happened when the water started moving over the ground? Most likely, it took things like rocks, soil, and leaves with it. This process of sediments moving from one place to another is called **erosion**. Here are some other examples of erosion:

- Glaciers carry boulders, pebbles, and soil from one place to another.
- Rivers wear down land to create valleys.
- Wind carves holes in rocks and builds mounds of sand in a desert.

SUPERCONTINENT

Earth's solid outer shell is broken into giant pieces, called **tectonic plates**. The plates are always on the move, but because they move very slowly, we can't feel it happening. They sometimes bump into each other and sometimes move apart. This constant movement causes major changes to land and water.

Millions of years ago, most of the continents on Earth were one piece of land—called a supercontinent—surrounded by ocean. That supercontinent was named **Pangaea**, which means "whole earth." Movement of the tectonic plates caused the land to break apart and drift away. Over time, Pangaea broke into seven pieces. They

PANGAEA

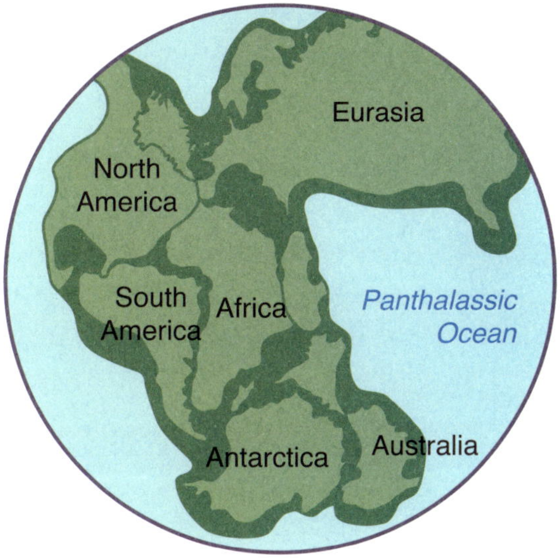

EARTH TODAY

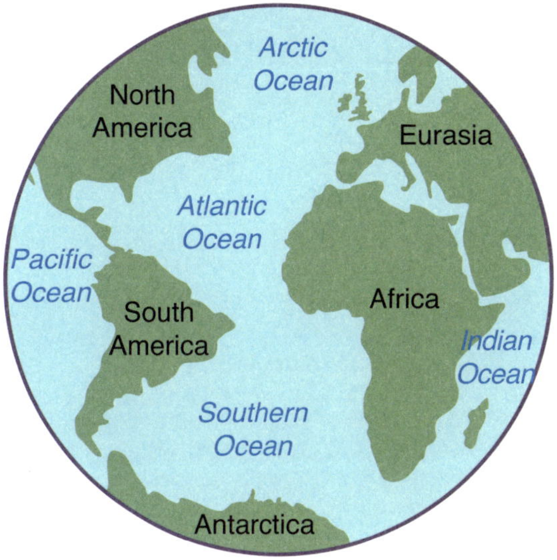

spread across the ocean and formed the continents we know today: Africa, Antarctica, Asia, Australia, Europe, North America, and South America.

But how do we know that Pangaea existed? Scientists noticed that the continents fit together like a jigsaw puzzle. For example, North America's eastern coastline matches the western coastline of Europe and South America's eastern coastline matches the western coastline of Africa. Geologists also found the same types of rocks and fossils on these continents, suggesting that they were joined together at one point in time and that the Atlantic Ocean had not always been there.

Scientists believe that there were other supercontinents before Pangaea. The continents have not stopped moving, but the movements are so slow that it would take millions of years to see another big change like that.

GEOLOGY IN ACTION

MOVING PLATES

What You'll Need:

POT

WATER

STOVETOP OR BURNER

7 PUZZLE PIECES

What to Do:

1. Fill a pot with water. Put the pot of water on a stovetop or burner. Do not turn the heat on yet.
2. Put the seven puzzle pieces together. Put the assembled pieces on top of the water in the pot. How are the pieces like Pangaea?
3. With an adult's help, turn the stovetop on. Let the water come to a boil. **Do not touch the hot water or the stovetop.** What happens to the puzzle pieces as the water gets hotter?
4. Turn off the stovetop. Talk about the activity with a friend or family member. How did the activity show what happened to Pangaea?

GET YOUR HANDS DIRTY

Go on a walk with an adult outside to look for signs of weathering and erosion. While you're on your walk, you may see cracked rocks or a path created by moving water. When you get home, draw what you saw. If you are keeping a geology journal, record your walk and what you found.

DIG DEEPER!

Many of the United States' National Parks were created by weathering and erosion. Use the National Park Service website (nps.gov) to learn more about our nation's parks. Which one would you like to visit most?

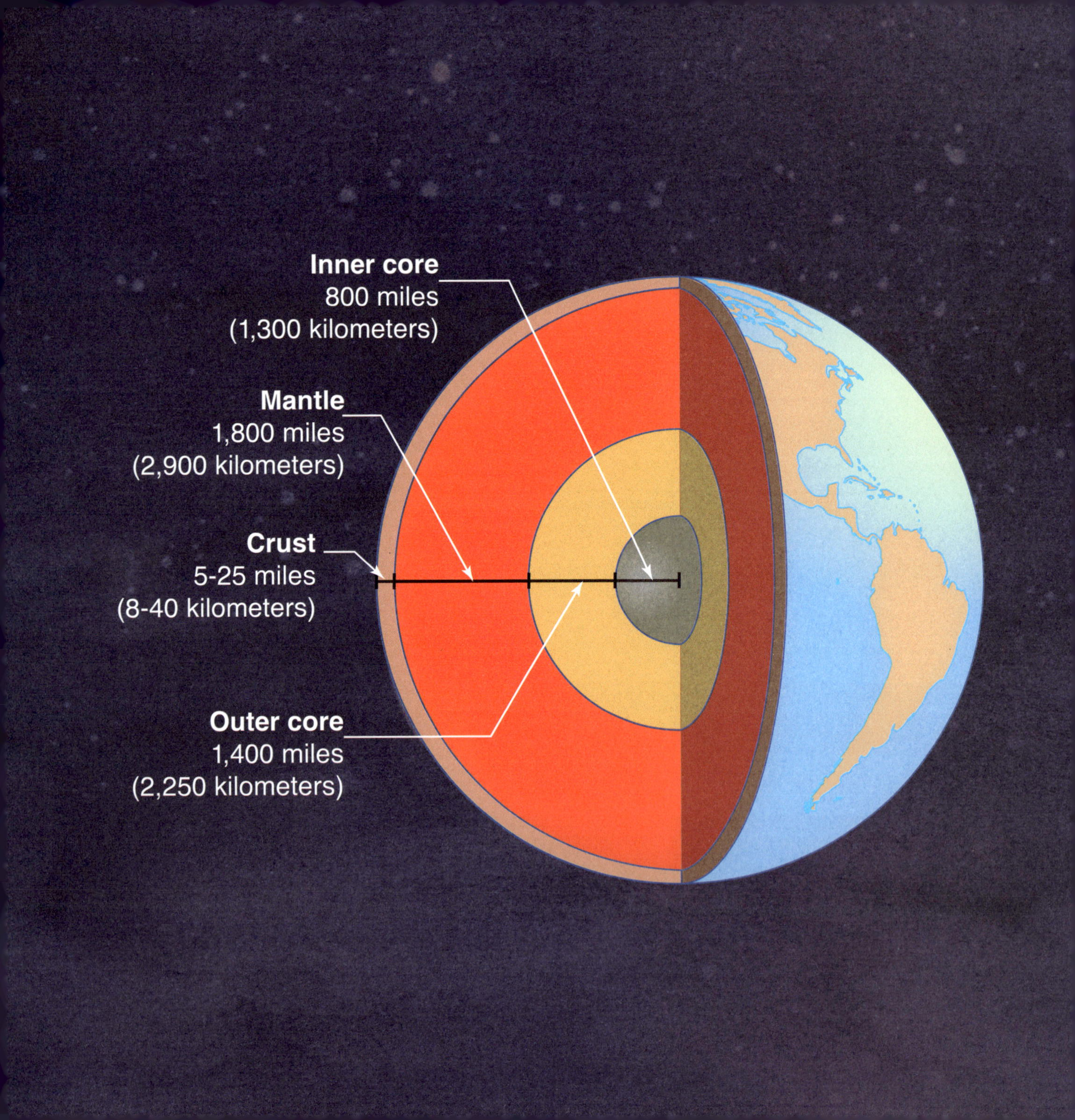
Inner core
800 miles
(1,300 kilometers)
Mantle
1,800 miles
(2,900 kilometers)
Crust
5-25 miles
(8-40 kilometers)
Outer core
1,400 miles
(2,250 kilometers)

Chapter Two

DIGGING DEEP

If you were to cut into a peach, you would find three layers: the pit, the soft flesh of the fruit, and the thin skin on the outside. The layers of Earth are similar. As we learned earlier, Earth has a dense center called a core, a large middle area called a mantle, and a thin layer on top called the crust. Each layer affects the one above it. This means that if something happens in the core, it causes changes in the mantle, and if something happens in the mantle, it causes changes in the crust. Actions deep within Earth can affect Earth's surface in big ways, even though the core is thousands of miles below the crust.

The Core

If you drop something heavy and dense into water, it sinks. Something similar happened when Earth began to take shape. The densest materials formed the deepest layer, and the lighter materials gathered on top. The core is divided into two parts: the inner core and the outer core. Both parts are made of an alloy, or combination, of two metals: iron and nickel, with a small amount of an unknown element, such as oxygen.

The inner core is about 4,000 miles below Earth's surface, which is about the same distance between New York and Germany. It is very, very hot—about as hot as the surface of the Sun. But the inner core doesn't melt. It stays solid because of the incredible pressure from the layers above it.

The outer core flows around the inner core. It's liquid because there is less pressure on it than on the inner core. This flow of the liquid creates an electric current. The current is what forms Earth's magnetic field and is why compasses always point north.

Because the core is so deep, it is difficult for scientists to study it. So how do they know anything about it? Because everything about Earth is connected, they can learn a lot from events like earthquakes. Geologists study the shock waves, or vibrations, caused by an earthquake to figure out the pressure, temperature, and chemistry of the inner and outer cores.

The Mantle

The mantle is Earth's largest layer. It is around 1,800 miles thick and makes up about 85% of the planet's volume. Like the core, the mantle is divided into two pieces: the lower mantle and the upper mantle. The upper mantle is between 200 and 360 miles thick. The lower mantle is far thicker.

Like the inner core, the lower mantle is very hot but does not melt because there is so much pressure on it. It is made of solid rock. The part of the lower mantle closest to the core can be as hot as 7,000 degrees Fahrenheit. That is 20 times hotter than the temperature needed to bake a cake!

A process called convection in the upper mantle can bring hot mantle rocks toward the surface and cause them to melt, forming **magma**, or molten rock. Magma can rise to the surface through cracks in the crust, then it is called **lava**. This is what happens when a volcano erupts. The part of the upper mantle closest to the crust is around 1,400 degrees Fahrenheit—more than 5,000 degrees Fahrenheit cooler than the deepest part of the mantle.

The Lithosphere

The **lithosphere** is the outer part of Earth. It is about 60 miles thick and is made of rocks from the crust and uppermost part of the mantle. The tectonic plates are fragments of the lithosphere. The seven major, or biggest, plates cover 95% of Earth. There are also several minor, or smaller, plates that cover the other 5%.

Tectonic plates are like rafts floating on top of the rock in the mantle. The plates are pulled along by dense slabs that sink into the deeper mantle at their edges. This is called plate tectonics. Plate tectonics has created many amazing features and events on Earth, like mountains, volcanoes, earthquakes, and even the shapes of the continents.

The place where two or more tectonic plates come together is called a **tectonic boundary**. There are three types of tectonic boundaries: convergent boundaries, divergent boundaries, and transform boundaries.

A convergent boundary is where two plates crash into each other. When they touch, one plate moves under the other, which pushes up the ground on the over-riding plate. This can form mountains. As the plates continue to push into each other, the mountains get higher. This process created the Himalayas, which have some of the highest peaks on Earth. Because the plates keep pushing into each other, these mountains are still growing—about one half inch each year!

A divergent boundary is where two plates move away from each other. When this happens, magma can ooze between the plates. The magma can rise to the surface and erupt to form a volcano. A divergent boundary can also be located underwater. Magma hardens when it touches water. This process creates underwater mountain ranges.

A transform boundary is where two plates slide and grind past each other. The grinding motion creates earthquakes. The San Andreas fault in California is a transform boundary between the North American and Pacific plates.

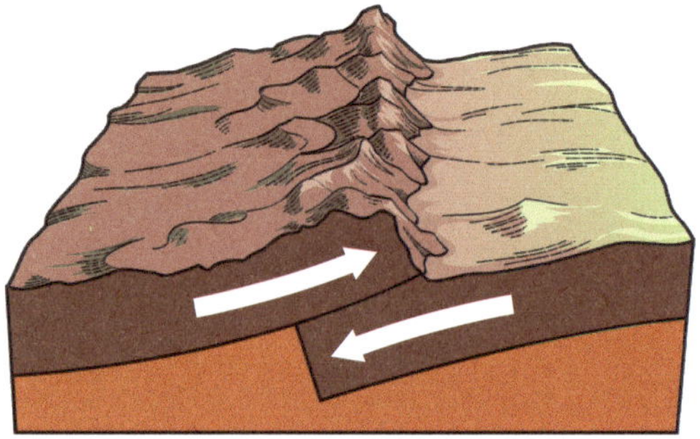

Convergent boundary: two tectonic plates moving toward each other, causing one plate to slide below the other

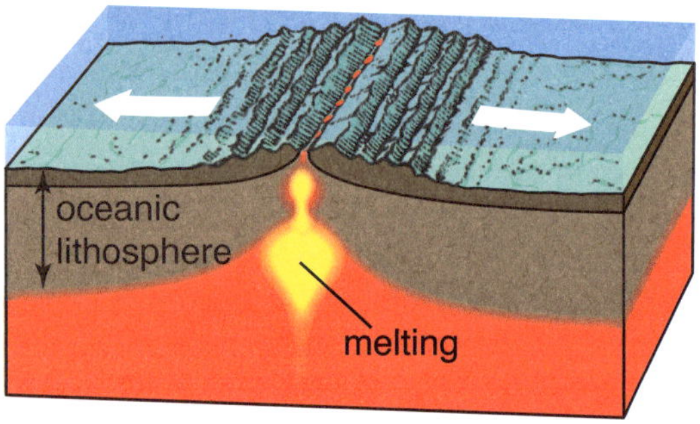

Divergent boundary: two tectonic plates moving away from each other, allowing magma to rise between

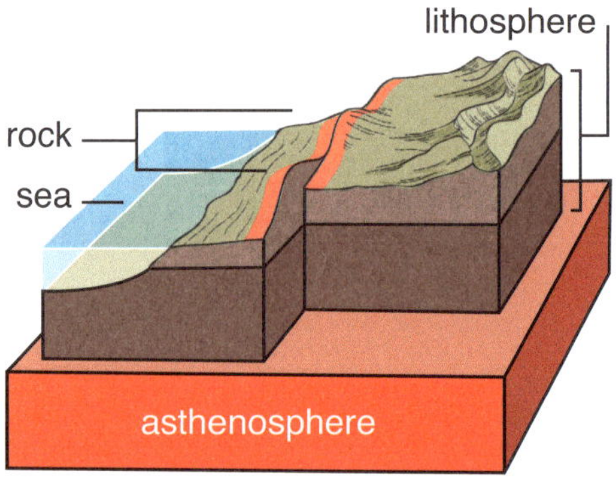

Transform boundary: two tectonic plates sliding and grinding past each other

The Crust

Earth's crust covers the entire planet. It may seem huge from the surface, but compared to the rest of the layers, it is very small. The crust makes up less than 1% of Earth.

More than half of Earth's crust is oceanic crust, which is the part found under the oceans. Oceanic crust is about four miles thick. Continental crust can be up to 45 miles thick. The average thickness is between 21 and 25 miles. It is made of different types of rocks called igneous, metamorphic, and sedimentary rocks, which we'll learn more about in chapter 6.

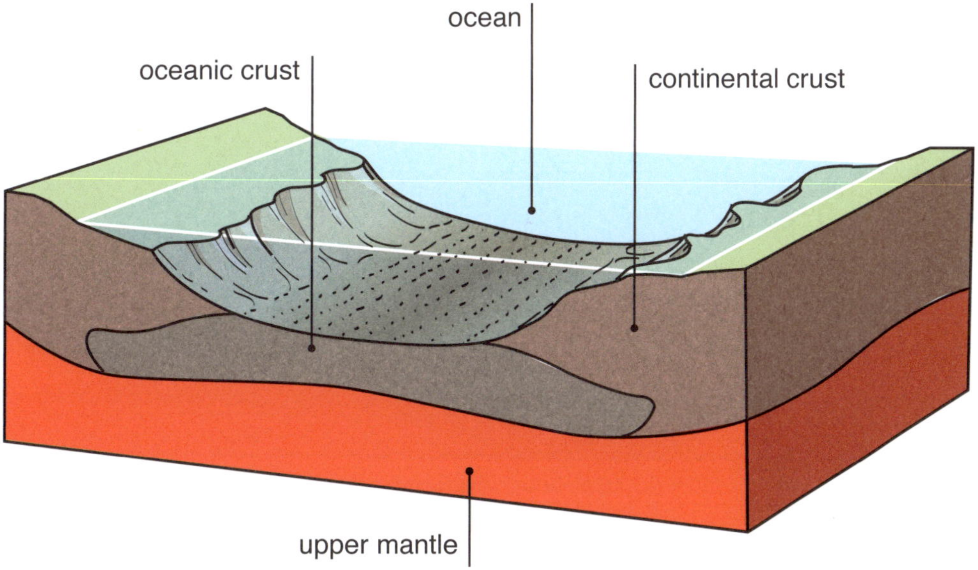

EARTH MODEL

What You'll Need:

MARBLE

TWO DIFFERENT COLORS OF CLAY

ALUMINUM FOIL

PLASTIC KNIFE

What to Do:

1. Let's create a model of Earth's layers. Begin with the marble. The marble represents the inner core.
2. Choose one color of clay. Wrap the clay around the marble. This represents the outer core.
3. Now let's make the mantle. Wrap the other color of clay around the outer core. Should you use more or less clay than you used for the outer core?
4. Carefully wrap a small piece of aluminum foil around the mantle. This represents the crust.
5. Carefully cut into your model with a plastic knife so you can see all the layers. Tell a friend or family member the name of each layer. Then, share one thing you know about each. What did you learn by creating this model?

GET YOUR HANDS DIRTY

With an adult's permission, find a place outside to dig. Dig through the top layer of soil. What do you notice? How does the soil change as you go deeper? Are there insects or worms? Fill the hole back in when you are finished. If you are keeping a geology journal, record your findings.

DIG DEEPER!

Spend some time learning more about tectonic boundaries. Have an adult help you use the Internet to research the different types. Find at least one land feature that was created by each boundary. Which are your favorite tectonic features?

Zebra Slot Canyon in Utah

Chapter Three

ON THE SURFACE

Earth looks different everywhere you go. Some places are flat, yet others have mountains. Some places are dry and brown, and others are wet and green.

The different features you see on Earth—like mountains, canyons, and deserts—are called **landforms**. Landforms are created by processes like shifting tectonic plates, erosion, and weathering. These processes can take millions of years. Let's check out some of the most common ones.

Mountains and Hills

Mountains are created where tectonic plates come together. The plates push into each other, which puts stress on the land. Over time, the land builds up. When the top of this rocky land pile is more than 1,000 feet taller than its surroundings, it is called a mountain. **Hills** are not as tall, less steep, and have a rounded top.

A single mountain is usually part of a mountain range. A **mountain range** is a line of connected mountains. Tectonic plates are huge. This means that the boundary between plates can be hundreds, or even thousands, of miles long. When the plates push together, mountains form along the entire boundary.

Mountains can be steep or have a gradual slope. They can be pointy or round on top. Mountains are formed many different ways. Let's check them out!

FOLD MOUNTAINS

Fold mountains are the most common type of mountains on Earth. The world's biggest mountain ranges are fold mountains, including the Himalayas, the Andes, and the Alps. These landforms are created when rocks and sediments are pushed up and folded on each other as the tectonic plates push together. To better understand how these mountains form, you can try this simple experiment. Cover a table with a tablecloth. Stand at one end of the table and ask a friend or family member to stand at the other end. At the same time, each of you should push the tablecloth toward the center of the table. You will notice that the tablecloth rises and folds to create a fold mountain range on your table!

Himalaya mountain range in Asia

The Rocky Mountains (left) stretch from British Columbia in Canada to New Mexico in the southwestern United States.

FAULT-BLOCK MOUNTAINS

Faults are cracks in Earth's crust, which may be large or small. Tectonic plate boundaries are a type of fault. Large blocks of rock can break when tectonic plates push into each other along a fault. This movement can force the blocks of rock upward and create fault-block mountains. The Sierra Nevada mountain range in California was formed this way.

Sierra Nevada mountain range in the western United States

VOLCANIC MOUNTAINS

As mentioned before, liquid rock, called magma, may form in Earth's mantle. As magma is less dense than the mantle and crust, it may rise through cracks in the lithosphere and erupt onto Earth's surface. The magma cools into lava and ash, then hardens. This process creates volcanic mountains. We will learn more about volcanoes in chapter 4.

Mount Fuji in Japan

DOME MOUNTAINS

Magma does not always reach Earth's surface. Sometimes it pushes up, but does not break through the crust, and then hardens before erupting. This process creates a dome mountain. Dome mountains aren't as steep and are more rounded than most other types of mountains.

Black Hills mountain range in South Dakota in the United States

MID-OCEAN RIDGE

Mountains are not only found on land. They are also under the oceans. In fact, the world's largest mountain range is mostly underwater. Called the mid-ocean ridge, it is 40,000 miles long and traces out the locations of divergent plate boundaries, as shown on the map. Most of it has never been seen by humans because 99% of the mountain range is beneath miles of ocean.

Valleys and Canyons

Valleys are the deep areas between hills and mountains. They can be different shapes and are formed in different ways.

RIVER VALLEYS

Make a V with two fingers. This is the shape of a river valley that has been formed by a river running between two mountains. Water erosion carries rocks and soil away, leaving behind a deep cut in the land. The sediments are dropped off in an ocean or lake at the end of the river.

A river valley in Nepal in South Asia

GLACIAL VALLEYS

Some valleys look more like the letter U. These valleys are usually formed by **glaciers**—large slow-moving sheets of ice. Glaciers are wider than rivers and move much more slowly. The ice catches and drags rocks and soil as it moves, leaving behind a wide glacial valley.

A glacial valley in the Swiss Alps of Switzerland

RIFT VALLEYS

Rift valleys are created by shifting tectonic plates. A rift, or break, in Earth's crust forms when plates move apart. This creates a flat valley where a block of the crust has dropped down, with steep walls on either side.

A rift valley in Iceland

CANYONS

A **canyon** is a very large, very deep valley with somewhat straight sides. Canyons are usually formed over millions of years as river water erodes and carves the land. There are different types of canyons. Slot canyons are formed by very fast-moving water. Box canyons have steep walls and an opening on only one side. Submarine canyons form on the ocean floor!

The largest canyon on Earth is a submarine canyon. Zhemchug Canyon is located on the bottom of the Bering Sea. It is two miles deep—far deeper than the famous Grand Canyon, which is just over one mile deep.

Zebra Slot Canyon in Utah

Plateaus

Plateaus are tall like mountains, but they are flat on top and have at least one steep side. Plateaus are some of the most common landforms on Earth and are found on all seven continents. These landforms can be found above and beneath the oceans.

Most plateaus are formed by tectonic activity where land is pushed upward due to tectonic plates colliding. They can also be formed from volcanic eruptions. After each eruption, a thin layer of lava cools and hardens. This gradually creates a smooth, flat surface. The largest plateau is the Tibetan Plateau, which is found in Asia. This plateau covers around one million square miles. It stretches through parts of China, India, and Nepal.

Waterberg Plateau in Namibia, Africa

Plains

Plains are large, mostly flat areas. A plain is formed when weathering and erosion flatten the land. Sometimes when a volcano erupts, the lava flows across the land. When it cools, it leaves a flat surface. This type of plain is called a lava field.

A plain in Tanzania, Africa

Continental plains look different depending on the location and climate:

Grasslands are plains found in drier places. They are covered in grass and have few trees.

Savannas are warm throughout the year. These plains have grass and trees that are spaced far apart.

Tundras form in very cold areas. The ground may be frozen, but small shrubs and moss still grow.

Floodplains are wet, flat areas next to rivers. These plains flood when the rivers rise.

Abyssal plains are large, flat regions at the bottom of the ocean.

Deserts

What do you think when you hear the word **desert**? Most likely, you imagine a large area of sand with a scorching hot Sun above. But did you know that not all deserts are hot? The hottest desert on Earth, the Sahara, can reach over 120 degrees Fahrenheit in the summer, but some deserts, like the Gobi, can get as cold as -40 degrees Fahrenheit during the winter. Not all deserts have sand, either. In fact, only 20% of the world's deserts are covered in sand.

Katpana Desert in Pakistan

The one thing all deserts have in common is that they are very dry. Most deserts get less than 10 inches of rain per year. Some see less than one inch in a year. Deserts also have few plants. The plants that live there have ways to survive with little water.

GLACIAL MOVEMENTS

What You'll Need:

16 OUNCES WHITE CRAFT GLUE

MIXING BOWL

1½ CUPS WARM WATER, DIVIDED

SPOON

PLASTIC CUP

1½ TEASPOONS BORAX POWDER

3 TO 5 BOOKS

COOKIE SHEET

ROCKS, SAND, SOIL, ETC.

What to Do:

1. Pour glue into a mixing bowl. Add 1 cup warm water and stir.
2. In a plastic cup, mix ½ cup warm water and 1½ teaspoons borax powder. Add to the glue mixture and stir. As the mixture becomes more solid, use your hands to roll it into a ball.
3. Stack several books. Set one end of the cookie sheet on top of the books to create an incline. Set the ball of dough on the cookie sheet at the top of the incline. Watch what happens. The movement you see is like the movement of a glacier.
4. Roll the dough into a ball again. Place things like rocks, sand, and soil on the cookie sheet. Set the ball of dough at the top of the cookie sheet again. This time, watch what happens to the rocks and soil. How does this model show what happens to the land as a glacier moves over it?

GET YOUR HANDS DIRTY

How many different landforms have you seen in your life? Can you see any hills, mountains, or bodies of water from your neighborhood? Every time you see a new landform, draw it or take a picture. How might they have been formed?

DIG DEEPER!

Mount Everest is very dangerous to climb. Few people have done it. Learn more about what it takes to climb Mount Everest. Read *The Top of the World: Climbing Mount Everest* by Steve Jenkins. What challenges do climbers face? What items would you need to climb this mighty mountain?

Mount Merapi in Indonesia

Chapter Four

VOLCANOES, EARTHQUAKES, AND MORE

In chapter 2, we learned that the mantle flows under Earth's crust. This helps the plates to move and push past each other. How does this affect us on the surface? In some very big ways! Things like volcanoes, earthquakes, and tsunamis are all caused by tectonic movement. Let's see how!

Volcanoes

Volcanoes are openings in Earth's crust where gas and magma escape from beneath the surface. Deep inside Earth, magma gathers together in magma chambers. The hot magma creates steam and other gases. Pressure builds up in the chamber. Eventually, the magma and gases are forced out of a crack in Earth's crust. This is a volcanic eruption. As soon as magma reaches Earth's surface, it is called lava. Volcanoes come in different shapes and sizes. There are four main types.

SHIELD VOLCANOES

Shield volcanoes look like a soldier's shield when viewed from above. They have gentle slopes, and some even look flat. These volcanoes form where superheated parts of the mantle rise toward Earth's surface, which creates big pools of magma beneath Earth's crust. Instead of exploding, shield volcanoes ooze runny lava during an eruption. This lava can travel very far, but it moves slowly. Many of Hawaii's volcanoes are shield volcanoes.

Mauna Loa in Hawaii

COMPOSITE VOLCANOES

Composite volcanoes have a tall, rocky cone shape. They form at convergent plate margins. The lava that comes out of a composite volcano is thick and sticky. Magma explodes out of a composite volcano through its top and cracks in its sides. The thick lava cannot travel very far. It piles up on the sides of the volcano while mixing with layers of ash and rock.

Mount Rainier in Washington state

CINDER CONE VOLCANOES

Cinder cone volcanoes also have a cone shape, with one hole at the very top. They often look like perfect circles from above. They may form in places where tectonic plates push together or move apart. The lava explodes out of the volcano, cools, and falls to the ground in little pieces of rock called cinders. The cinders pile up all around the hole to form the cone.

Kostal Cone in British Columbia, Canada

LAVA DOME VOLCANOES

Lava dome volcanoes release lava that is too thick to travel very far at all. The lava piles up around the opening as it erupts. Then, it hardens to create a dome shape. The dome grows larger and larger with each eruption. Lava domes can be explosive, but usually they do not have enough gas or pressure to erupt with an explosion.

Mount Merapi in Indonesia

ACTIVE, DORMANT, AND EXTINCT VOLCANOES

Scientists also classify volcanoes based on when they last erupted. There are three groups:

Active volcanoes have erupted any time in the last 10,000 years. An eruption could be lava or gas.

Dormant volcanoes have not erupted in the last 10,000 years, but scientists believe that they could become active again.

Extinct volcanoes are not active, and scientists do not believe that they will become active again.

Today, there are about 1,500 active volcanoes around the world. Scientists believe that at any given time, 40 to 50 volcanoes on Earth are erupting. There is no way to know exactly when a volcano will erupt. Geologists watch for certain warning signs to make the best predictions possible. Earthquakes near a volcano, changes to the shape of the volcano, or more gas coming out of the volcano are all warning signs that a volcano might erupt soon.

THE RING OF FIRE

The Ring of Fire is a 25,000-mile-long string of volcanoes around the Pacific Ocean. It has more than 450 volcanoes—about 75% of all volcanoes on Earth. Several of them erupt every day! The ring is formed where several tectonic plates meet several other plates.

Most of the tectonic plates around the Ring of Fire are pushing up against each other. Often, one tectonic plate will slide under another tectonic plate as they push together. This is called a subduction zone. Heat and pressure cause the plate that slides underneath to melt. The melted magma and plate movement causes volcanic eruptions and earthquakes. Most of the world's earthquakes happen along the Ring of Fire. They are often the most extreme in the world.

Earthquakes

Earthquakes happen at edges of tectonic plates that are rough and rocky. When the plates move, they do not slide smoothly. Instead, they bend and twist as they rub together. This movement sends vibrations through the Earth. As the vibrations move toward the surface, they shake the ground. We feel it as an earthquake.

Aftermath of the San Francisco earthquake of 1906

Some earthquakes are so small that we can't feel them. Other earthquakes are so powerful they can destroy entire cities. Seismologists are people who study earthquakes. Until recently, they used the Richter scale to decide how strong an earthquake is. The Richter scale gives an earthquake a number, called a magnitude, from 1 to 9 that is related to the strength of its vibrations. An earthquake with a magnitude of 1 is not noticeable. An earthquake with a magnitude of 9 causes the most damage. The standard scale used by seismologists today is called the moment-magnitude scale, which is more accurate.

Scientists use a seismograph to measure an earthquake's strength on a magnitude scale. A seismograph is a large machine that sits on the ground. When the land shakes, the seismograph draws a line that shows how strong and fast the earthquake is. The larger the wiggly lines, the stronger the earthquake.

Tsunamis

Have you ever been to the beach? How tall were the waves you saw? Most likely, they were only a few feet tall. Tsunami waves take things to a whole new level. They can be as tall as a 10-story building!

A **tsunami** is a bunch of waves that are created when the ground underneath the ocean moves. This can happen with a volcano or an earthquake. The movement of tectonic plates under the water makes waves that travel out in all directions. Sometimes, these waves are small and go unnoticed. However, after a large underwater volcano or earthquake, tsunami waves can be anywhere from 10 to 100 feet tall.

A tsunami can travel over 500 miles per hour. That's almost as fast as an airplane! Because of their speed, tsunami waves can cause a lot of trouble if they make it to land. The water rushes onto the coast, making it difficult for people to get out of the way. They can flood cities in minutes.

A very large tsunami happened in 2004 when a magnitude 9.1 earthquake shook the ocean floor near Indonesia. Thirty-foot waves flooded nearby land within a few hours of the earthquake. It destroyed many homes and towns.

Tsunami wave

VOLCANO IN A BOWL

What You'll Need:

- **3 CRAYONS**
- **STOVETOP-SAFE GLASS CONTAINER**
- **SAND**
- **WATER**
- **STOVETOP OR BURNER**

What to Do:

1. Remove any paper from the crayons. Sit the crayons in the center of the glass container.
2. Pour sand on the crayons. Make sure the crayons are completely covered.
3. Pour water into the glass container. There should be at least 1 inch of water above the sand.
4. With the help of an adult, sit the glass container on the stovetop. Turn the stovetop to medium-high. Do not touch the glass container or stovetop.
5. It will likely take several minutes for something to happen. What do you see? Talk about how the experiment showed how a volcano erupts. If you are keeping a geology journal, record what you saw and learned. Make sure you draw a picture of your experiment!

DID YOU KNOW?

Earth isn't the only place you can find volcanoes. Scientists have found them on many planets and moons. The largest volcano that we know of is on another planet. Olympus Mons is a giant volcano on Mars. It is more than 13 miles tall!

DIG DEEPER!

Earthquakes happen every day. Check out the United States Geological Survey's Latest Earthquakes webpage (earthquake.usgs.gov/earthquakes/map/) to find out when and where the most recent earthquakes happened. Where was the earthquake closest to you?

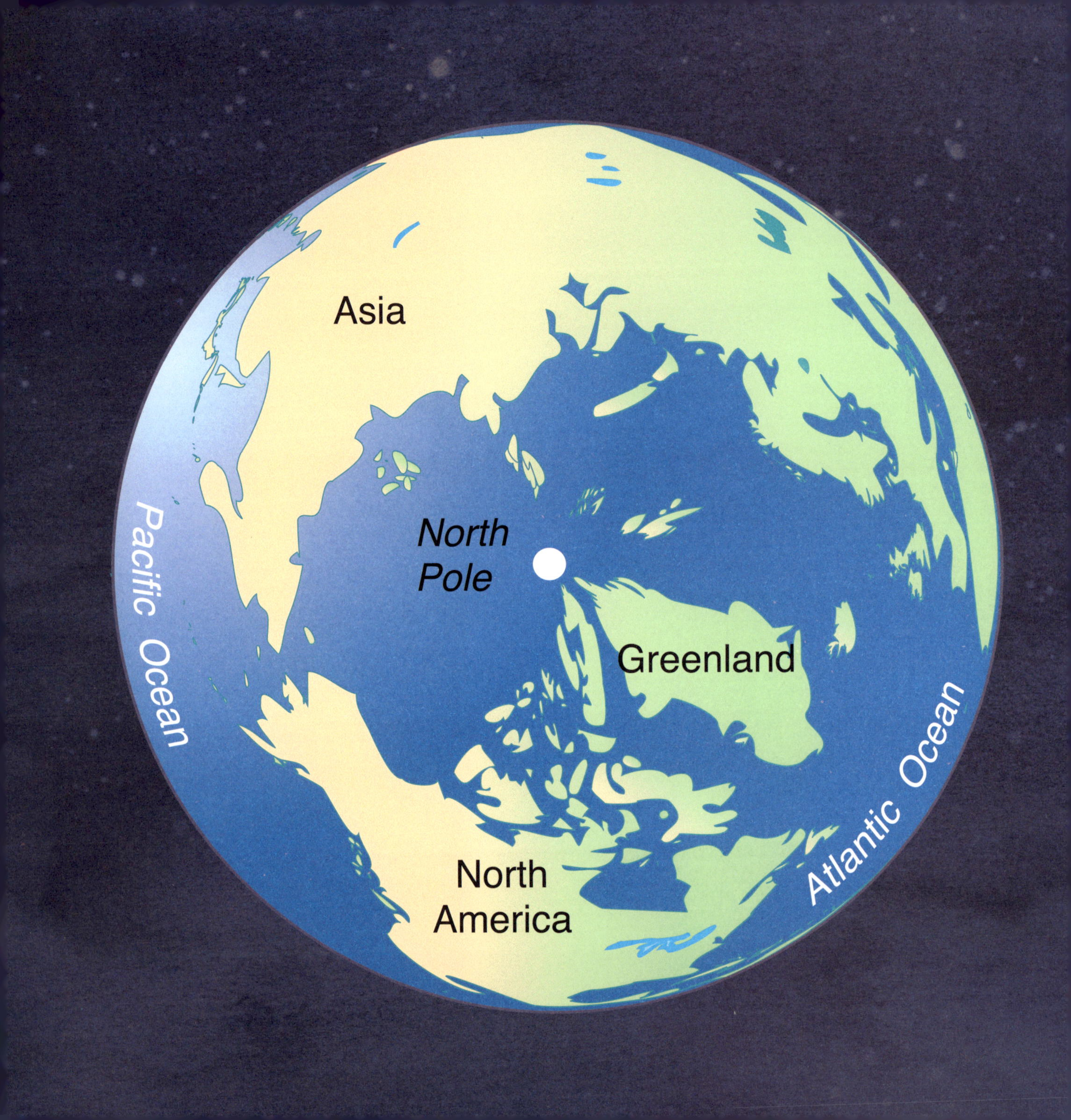
Asia
North Pole
Pacific Ocean
Greenland
Atlantic Ocean
North America

Chapter Five

EARTH'S WATER

If you look at a globe, you will see more blue than green. The blue areas are Earth's oceans, lakes, and rivers. They hold the water that all living things need to survive. Earth's water is all connected. Rivers and streams flow into the oceans and all the oceans flow into each other. They form one giant body of water that we call the World Ocean.

Oceans

If Earth was formed from dust, you might be wondering where all the water came from. The oceans formed billions of years ago, very soon after Earth itself formed. The dust and rocks that shaped Earth had water in them. Earth's extreme heat caused the water to turn into a gas called **water vapor**. The vapor formed clouds. Over time, the clouds cooled. Rain fell from the clouds, and the water collected on Earth to form the oceans.

Oceans are large bodies of salt water that cover most of Earth's surface. There are five ocean areas: the Pacific Ocean, Atlantic Ocean, Indian Ocean, Southern Ocean, and Arctic Ocean. Which is closest to where you live?

PACIFIC OCEAN

The Pacific Ocean is located between Asia and Australia and North and South America. It is the largest ocean on Earth—more than 60 million square miles. That is more space than all the land on Earth put together! The Pacific is also the deepest ocean. Its deepest point is nearly 36,000 feet under the water. That's how high airplanes fly in the sky! On average, the Pacific Ocean is about two miles deep. Volcanoes, earthquakes, and tsunamis are common in the Pacific Ocean because the Ring of Fire is located there.

ATLANTIC OCEAN

The next largest ocean is the Atlantic Ocean. Even though it is big, it is only half the size of the Pacific Ocean and is located between North and South America and Europe and Africa. The Atlantic is home to millions of fish and marine animals, so it is very popular for fishing. It is not as deep as the Pacific, but the deepest spot in the Atlantic Ocean is almost deep enough to cover Mt. Everest (the tallest mountain on Earth) in water.

INDIAN OCEAN

The Indian Ocean is Earth's third-largest ocean. It covers about 28 million square miles between Africa, Asia, and Australia. The deepest part of the Indian Ocean is about 24,000 feet below sea level. It is the warmest ocean on Earth, which makes it hard for many fish and marine animals to survive.

SOUTHERN OCEAN

The fourth-largest ocean is the Southern Ocean. The Southern Ocean is sometimes called the Antarctic Ocean because it is located around Antarctica. Because of its location on the globe, the Southern Ocean touches the Indian Ocean, Atlantic Ocean, and Pacific Ocean. It is much smaller than the oceans it touches. You could almost fit three Southern Oceans inside the Indian Ocean! The water is very cold—about 28 degrees Fahrenheit—so giant icebergs are a common sight!

ARCTIC OCEAN

The Arctic Ocean is the smallest ocean. It is located around the North Pole, above North America, Europe, and Asia. It is very cold that far north, so the Arctic Ocean is often covered in sheets of ice. Earth's average temperature has been rising in recent years, so many ice sheets are melting. This is making the water level rise all over the globe.

THE OCEAN FLOOR

If you could travel to the ocean floor, you would find a whole world with mountains, valleys, and volcanoes. Underwater mountains are formed the same way as mountains on the surface. Tectonic plates rub up against each other and push up rock. Trenches are deep valleys in the ocean floor. These underwater valleys are the deepest spots in the ocean. Trenches are formed where tectonic plates converge.

Rivers, Streams, and Lakes

Maggia River in Switzerland

The oceans' salt water makes up nearly 97% of Earth's water. The rest of Earth's water is fresh water, or water without salt. Most of the fresh water on Earth is frozen in ice and glaciers. Some is in groundwater, water that is under the ground in soil and rocks. A small amount of Earth's fresh water flows in rivers, streams, and lakes.

A river is a body of moving water that begins at an uphill source. The source is usually a melting glacier, melting snow, a lake, rain, or groundwater in the mountains. The water flows downhill until it reaches the end of the river, which is called the mouth. The mouth empties into a larger water source such as an ocean or a lake.

A stream is very similar to a river, but is much smaller. Often a stream is a tributary. This means that the stream joins a larger river before it reaches an ocean or lake.

Lakes are bodies of water surrounded by land. Rivers, streams, melted glaciers, and rainwater all flow into lakes. The largest freshwater lake in the world is Lake Superior, located between the United States and Canada.

Some lakes, like the Dead Sea and the Great Salt Lake, contain salt water instead of fresh water.

The Water Cycle

Did you know that when you take a sip of water you are sipping something that is billions of years old? That water has been sipped by many people and animals before you. All the water on Earth goes through a three-part cycle that happens over and over again. This **water cycle** causes rain, fills rivers and oceans, and makes our drinking water.

The first stage of the water cycle is **evaporation**. This is when some of the water in oceans, lakes, rivers, and streams becomes water vapor and rises into the air. Water vapor is invisible.

The second stage of the water cycle is **condensation**. As water vapor rises, it cools. Once it is cold enough, the water vapor turns back into tiny drops of liquid water. These tiny drops of water form clouds.

The final stage of the water cycle is **precipitation**. When the clouds become too heavy, the drops of water fall back down to Earth. If the temperature is above freezing, the drops fall as rain. When the temperature is below freezing, the water freezes and falls as snow, sleet, or hail. The water collects in oceans, lakes, or rivers and stays there until it evaporates back into the air. Then the water cycle starts all over again.

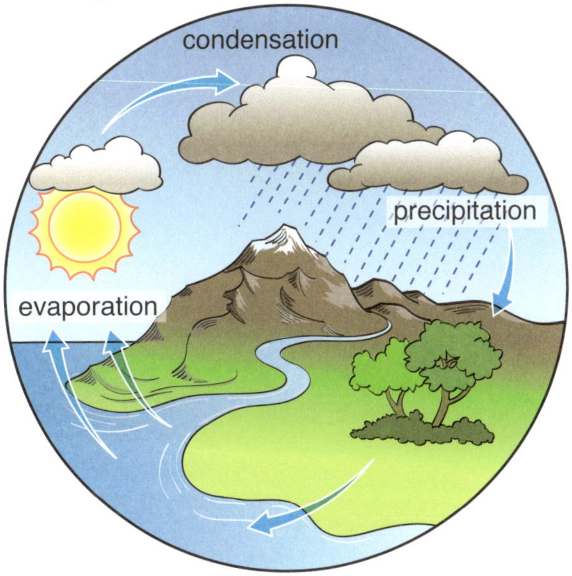

CLOUD IN A JAR

What You'll Need:

WATER

POT

STOVETOP OR BURNER

CLEAR GLASS JAR WITH LID

ICE

AEROSOL HAIRSPRAY

What to Do:

1. With the help of an adult, bring water to a boil in a pot on a stovetop.
2. Ask an adult to pour the hot water into the jar. The jar should be about half full.
3. With an adult's help, hold the rim of the jar and gently swirl the water around. This will heat the sides of the glass. Be careful not to splash the hot water or touch the hot parts of the jar.
4. Place the lid upside-down on top of the jar. Fill the lid with ice. Let the ice sit on top of the jar for 1 minute.
5. After 1 minute, lift the lid and pump a few sprays of hairspray into the jar. Put the ice-filled lid back on top of the jar. Do this quickly so that the hot air does not escape.
6. Watch what happens in the jar. What do you think causes this to happen? What does it teach you about the water cycle?

DID YOU KNOW?

Lake Superior is gigantic. The water it holds could cover all North and South America in a foot of water. Where does all the water come from? More than 300 rivers and streams end their journey at this Great Lake.

DIG DEEPER!

Read *A Drop Around the World* by Barbara Shaw McKinney. It explains the journey a drop of water takes through clouds, ice, puddles, and rain. As you read, point out the three stages of the water cycle.

Ski huts on Mount Ruapehu in New Zealand

Chapter Six

ROCKS AND MINERALS

Earth is made mostly of rocks, and rocks are made of minerals. Both rocks and minerals are important to people, plants, and animals. We use them to build houses and schools, roads and sidewalks, and even forks, spoons, and cooking pots. Minerals in the soil feed plants and help our bodies build strong bones and teeth. But what exactly are rocks and minerals, and how do you tell them apart?

What Is a Mineral?

Minerals are all around you. They are a part of your daily life. The salt you sprinkle on food is a mineral. Even diamonds and gold used to make jewelry are minerals.

You might wonder how two things as different as salt and gold could both be minerals. Scientists say that for something to be a mineral, all five of the following facts need to be true. As you read, think about salt and gold. Are these statements all true for both?

Minerals are naturally occurring.

This means that minerals are not made by people. They are created in nature.

Minerals are solids.

They are not liquids or gases.

Minerals have a definite chemical composition.

This sounds complicated, but it just means that each mineral is made of certain chemical ***elements***. *There are 118 elements. A mineral can contain one or more elements. Carbon, oxygen, and copper are all elements.*

Minerals are inorganic.

Inorganic means that minerals are not made by a living thing. A plant is not inorganic because it is made from the seed of another plant. Animals are not inorganic because they are made by their parents. Minerals are not made from plants or animals. Because they are not made by another living thing, minerals are inorganic.

Minerals have an ordered internal structure.

You've learned that minerals are made of chemical elements. An ordered internal structure means that the chemical elements for each mineral are organized the same way.

There are more than 4,000 different minerals. Each mineral forms a different way. Some form on Earth's surface when lava cools. Others form inside

Earth when magma cools. Minerals also form in the extreme heat and pressure below Earth's crust.

Check out this chart to learn about some everyday minerals and how they are used by humans. Do you have any in your home?

MINERAL	HOW IT IS USED
 DIAMOND	Diamonds are used to make jewelry. They are also used to make some tools like drill bits, polishing tools, and saws.
 QUARTZ	Quartz is used to make glass and jewelry. It is also used to make some computer parts.
 TALC	Talc is often used in makeup, deodorant, and baby powder. It is also used to make some plastics and paints.

MINERAL	HOW IT IS USED
 CALCITE	Calcite is one of the most common minerals. It is used to make construction materials. Calcite is also used in some medications.
 SILVER	Silver is used to make many objects. Some objects include mirrors, eating utensils, and money. Silver is a great electrical conductor. This means it can be used for electrical wiring.

What Is a Rock?

You don't have to look far to find rocks. They are everywhere. But what is a rock? Every type of rock is made of minerals that have become packed together. A rock can be made of one mineral or many. The types of minerals a rock is made of are what give the rock its appearance.

Rocks are divided into three categories, or groups: igneous, metamorphic, and sedimentary. The categories tell us how the rock formed, what it looks like, and where it can be found. Understanding each type can help you identify the rocks you find in your daily life or while on vacation.

IGNEOUS ROCKS

When a volcano erupts, lava flows onto Earth's surface. Have you have ever wondered what happens to the lava? It doesn't stay liquid forever. Eventually, the lava cools and hardens. When lava cools and hardens, it forms an **igneous rock**.

Igneous rocks formed by lava *outside* Earth are called extrusive igneous rocks. Some igneous rocks form when magma cools and hardens *inside* Earth. These rocks are called intrusive igneous rocks. Whether extrusive or intrusive, igneous rocks are the most common rock type in Earth's crust.

Take a look at some of the most common types of igneous rocks and how they are used. How are the intrusive rocks similar? How are the extrusive rocks similar?

GRANITE

Granite is an intrusive igneous rock. It often has large crystals scattered throughout. This stone is used to make countertops and floors. You may even have granite in your home.

OBSIDIAN

Obsidian is an extrusive igneous rock. Some people call it volcanic glass because it is smooth and shiny and has sharp edges. Obsidian is sometimes used to make extremely sharp knives and tools.

PUMICE

Pumice is an extrusive igneous rock. Pumice has tiny holes all over it. The holes form because gas gets trapped inside the lava as it cools. Some people use pumice stones when bathing. They are rough and are good at removing dead skin.

GABBRO

Gabbro is an intrusive igneous rock. It is usually a dark color. Many construction materials contain gabbro. It is used to make roads, bridges, and even countertops.

SEDIMENTARY ROCKS

A clue to how sedimentary rocks form is in their name. **Sedimentary rocks** are formed from . . . sediments! Erosion moves particles of minerals and plants into low-lying areas like lakes, oceans, and valleys. Over millions of years, the sediments pile up on top of each other. The weight of all that material presses the sediments together and eventually forms sedimentary rocks. This type of rock is often found near water, but some can also be found in dry places.

Here are some of the most common types of sedimentary rocks. Each type of sedimentary rock is formed from different sediments.

COAL

Coal is a black stone mostly formed from ancient dead plants. It burns and is often used as fuel to make electricity. It provides about 25% of the world's energy.

LIMESTONE

Limestone is usually formed underwater from shells, coral, and algae. Limestone was used to build the Great Pyramid in Egypt and many ancient and modern buildings. Today it is used to make many things, including concrete, toothpaste, paint, and baking soda.

SANDSTONE

Sandstone is formed from different types of sand. It can be red, tan, brown, yellow, or gray. Sandstone is used to make house tiles, columns, and walkways.

CONGLOMERATE

Conglomerate forms from pieces of other rocks, sand, and mud that are cemented together. Conglomerate is sometimes used for buildings. It is often crushed for road-paving.

METAMORPHIC ROCKS

Most **metamorphic rocks** form deep within Earth. These are already-formed rocks that change into new rocks because of the high heat and pressure under the ground. The heat usually comes from nearby magma. Metamorphic rocks can form from igneous, sedimentary, or even other metamorphic rocks.

Here are some of the most common types of metamorphic rocks.

GNEISS

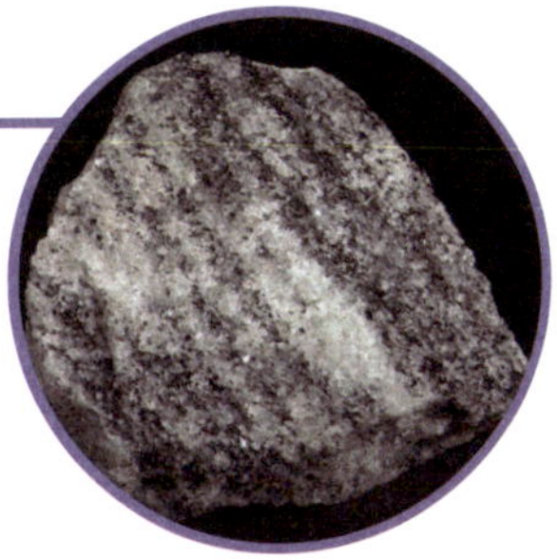

Gneiss looks like granite, but with stripy patterns instead of speckles. It is used to make monuments and gravestones. It is also used in the construction of some stone buildings and for floors and countertops.

MARBLE

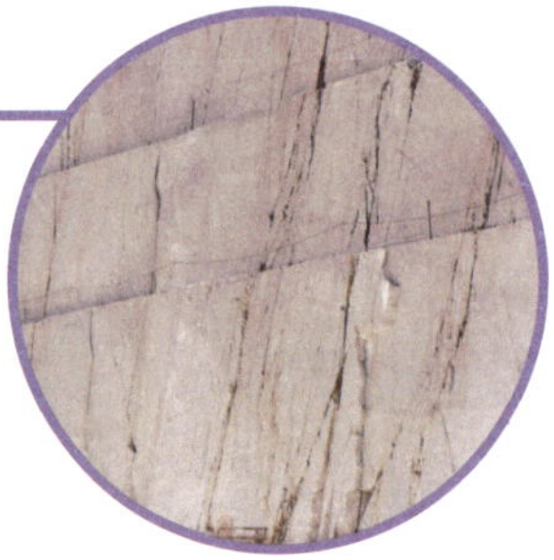

Marble was limestone before becoming metamorphic rock. It has been used to build some of the United States' most popular buildings and monuments. The Supreme Court Building, Washington Monument, and Lincoln Memorial are all made of marble.

SLATE

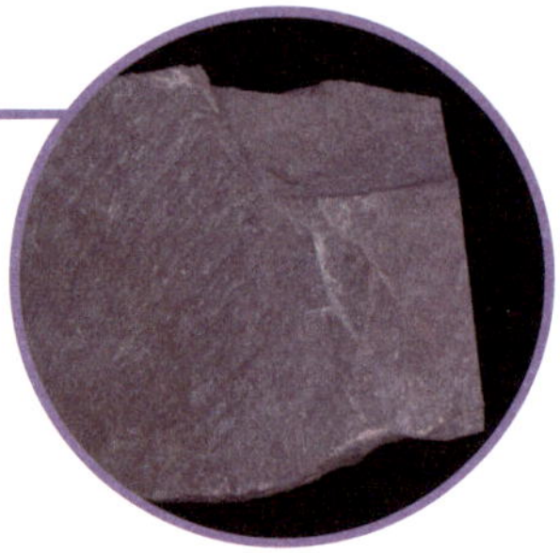

Slate was a sedimentary rock called shale before becoming metamorphic rock. Slate is usually gray, but can be many colors. It splits easily into thin pieces. Slate is used for roof tiles and floors. Blackboards for classrooms are also made from it.

QUARTZITE

Quartzite was sandstone before becoming a metamorphic rock. It is usually white, pink, or gray. It is made of quartz, which is where it gets its name. Quartzite is often used to make countertops and tiles.

Fossils

Sometimes, sedimentary rocks have layers. The layers are the different sediments that piled up over time. Some sedimentary rocks contain fossils. **Fossils** are the remains of ancient plants or animals. When a plant or animal dies, sediments collect around the remains. Over time, the plant or animal decays, but it leaves something behind—an imprint. This is a fossil. Sometimes actual shells or bones become fossils.

Scientists use fossils to learn about Earth. Fossils found in rock layers show what plants and animals were around when the rocks were formed. Scientists use technology to learn how old a fossil is. When they learn how old a fossil is, they also learn the age of the rocks where the fossil was found. This process is called dating.

The Rock Cycle

Let's say you find a rock. You pick it up and turn it around in your hands. You notice it has layers and the imprint of a shell on one side. What do you know about this rock? These features tell you it is a sedimentary rock, but this rock hasn't always been that way. At one time, it may have been an igneous or a metamorphic rock. Like all the water on Earth, all the rocks on Earth are changing as they move through a constant cycle. This cycle is called the **rock cycle**.

Here is an example of the rock cycle: Lava spills from a volcano and cools on Earth's surface. An igneous rock forms. Over millions of years, rain and wind break the rock into sediments. Erosion sweeps the sediments into the bottom of a lake. Over many more millions of years, the sediments press together and form a sedimentary rock. More time passes and tectonic plate movements push the rock down inside Earth. Heat and pressure change the rock into metamorphic rock. Eventually, Earth's heat melts the rock into magma. As the magma rises to Earth's surface, the rock cycle starts over again with a new volcanic eruption.

Look at the diagram to learn more about the rock cycle. Does the rock cycle constantly move in the same direction?

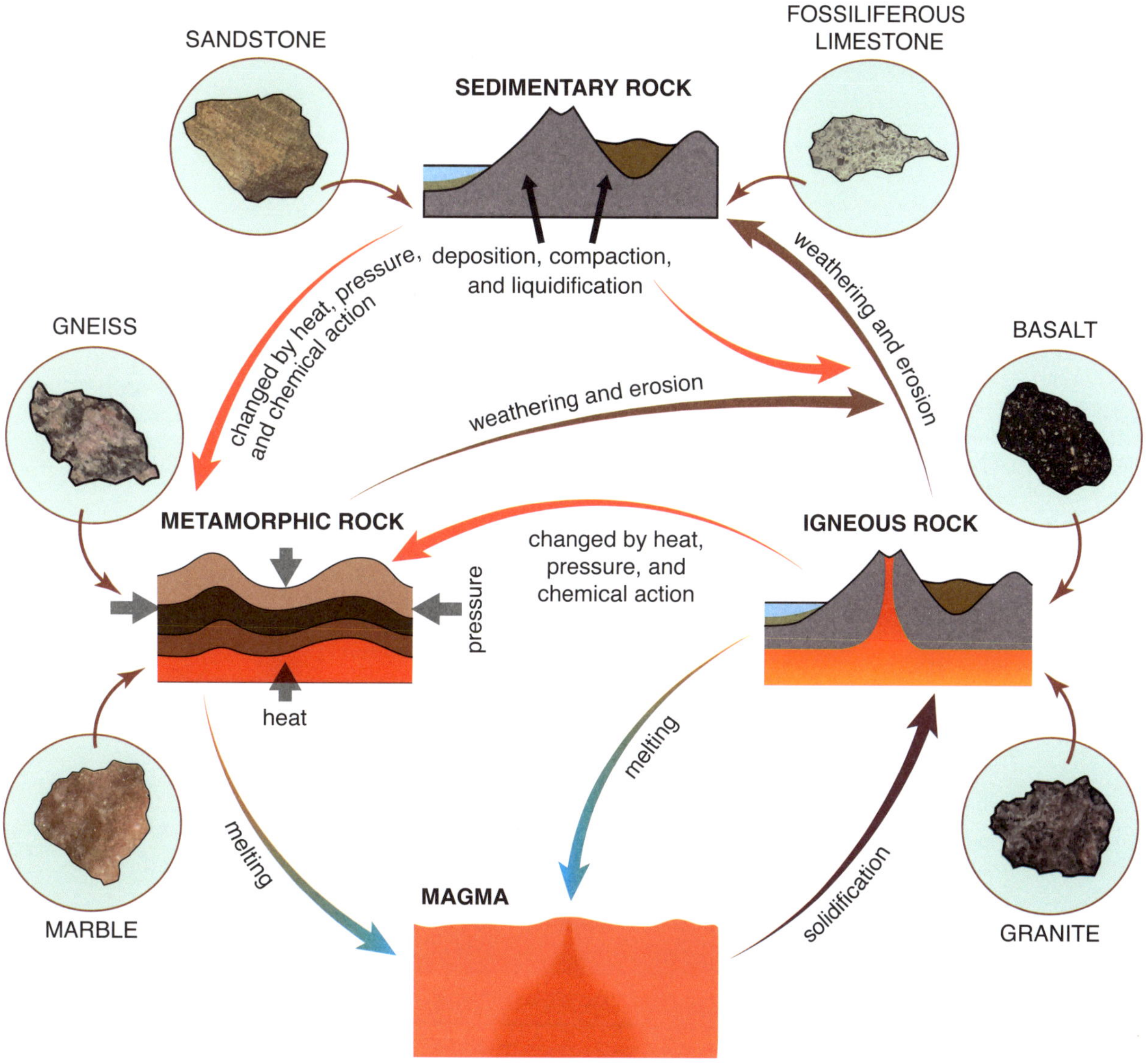
SANDSTONE
SEDIMENTARY ROCK
FOSSILIFEROUS LIMESTONE
changed by heat, pressure, and chemical action
deposition, compaction, and liquidification
weathering and erosion
GNEISS
BASALT
weathering and erosion
METAMORPHIC ROCK
IGNEOUS ROCK
changed by heat, pressure, and chemical action
pressure
heat
melting
melting
MARBLE
MAGMA
solidification
GRANITE

CANDY ROCK CYCLE

What You'll Need:

3 STARBURST CANDIES (USE DIFFERENT COLORS)

PAPER PLATE

MICROWAVE-SAFE BOWL

MICROWAVE

What to Do:

1. Unwrap the candies. Stack the 3 candies on top of each other on a paper plate. Put your hand on top of the pile and press down. What type of candy rock did you form?
2. Next, put the candy inside your hand. Make a tight fist around the candy. Hold it tightly in your fist for at least 1 minute. Then, roll the candy between your hands. What type of candy rock did you form?
3. Put the candy in a microwave-safe bowl. With the help of an adult, heat the candy in the microwave for 15 seconds. Remove the candy and let it cool. What type of candy rock did you form?
4. Discuss how you modeled the rock cycle with a friend or family member.

GET YOUR HANDS DIRTY

Go outside with an adult and find five different rocks. Examine your rocks. Do they give you clues about what type of rocks they are? If you are keeping a geology journal, draw pictures of your rocks and note where you found them.

DIG DEEPER!

Read *Everything Rocks and Minerals* by Steve Tomecek. Find two rocks and two minerals that you knew nothing about before reading the book. What made you notice them? Which is your favorite?

Rock collection

Chapter Seven

A GUIDE FOR ROCK HOUNDS

Has learning about geology made you excited about rocks? Then, it's time to start building your own rock and mineral collection! Here are some easy ways to find, test, and identify rocks and minerals for your collection. Anyone can be a rock hound! Let's get started.

The Neighborhood Geologist

Many geologists travel the world. They use expensive equipment to study rocks. But, you do not need to travel far or have fancy tools to be a junior geologist. You can search for rocks and minerals almost anywhere, even your own neighborhood or town!

Often, rocks form in and around water. When you are searching, be sure to check near water features such as oceans, lakes, rivers, streams, and creeks.

If there are not any water features near you, that's okay. There are plenty of other places to look for rocks in nature. It can be as simple as searching your local park or playground. You can also look for rocks while on a hike through the woods or in the mountains. Keep in mind that you are more likely to find rocks and minerals in places like hills, mountains, beaches, and valleys than flatlands such as plains. Make sure to ask permission before moving any rocks that are not on your property.

When searching for rocks and minerals, safety is important. Follow these tips to stay safe:

- An adult should always know where you are.
- Always wear closed-toe shoes to protect your feet.
- Wear gloves to protect your hands.
- Watch where you step to prevent slipping and falling. It is especially important to be careful around water and cliffs.

Gather Your Tools

As you collect rocks, store them in a small container or bag. If you are keeping a geology journal, take it with you. Write down where you find each rock after you pick it up. Once you've collected a bunch of rocks, it is time to head home. The rocks and minerals that you collected are called **specimens**. A specimen is a sample of something that is used for testing.

Next, gather the materials needed to study your specimens. You will need the following materials to clean, study, and identify the specimens in your collection:

- toothbrush
- water
- paper towels
- paper and pencil
- magnifying glass
- black and/or white porcelain streak plate or unglazed tile
- steel nail
- penny

If you have a geology journal, keep it handy. A good geologist always takes notes and draws pictures as they work. This helps them remember the things they see.

Rock or Mineral?

It's time to get to work! The first thing you should do is clean your specimens. Because rocks and minerals are formed in the ground and found outside, they are often covered in dirt. It is important to be able to see what the specimens really look like in order to identify what they are.

To clean your collection, use water and a toothbrush. Dip the toothbrush in the water. Then, gently scrub all the specimens until there is no dirt left on them. After the specimens are clean, leave them on a paper towel to dry.

Next, number the specimens so that you can keep track of each one that you study. Arrange your specimens in rows on a piece of paper. Then, write a number next to each one. When you examine your specimens, only pick up one at a time. When you are finished studying a sample, put it back next to the correct number on the paper.

The first step in identifying each specimen is to decide whether it is a rock or a mineral. To do this, pick up a specimen and look it over closely. It is best to use a magnifying glass to get a closer look. Rocks and minerals each have unique features. Use the chart to help you determine whether each specimen is a rock or a mineral. Keep in mind that not all rocks will have all the features of a rock and not all minerals will have all the features of a mineral.

Record what you find in your geology journal, if you are keeping one. It is a good idea to write a number on the top of each page in your journal. The numbers at the top of each page will match the numbers next to your specimens. For example, write about specimen 1 on page 1 of your journal. Then, record all the information you discover during your study.

ROCKS	MINERALS
Most rocks are made of more than one type of mineral.	Minerals have a fixed composition. Only one material is visible throughout.
Often, you can see different materials in the specimen.	Most minerals are only one color throughout.
Most rocks have different colors or shades of colors throughout.	Minerals do not have fossils.
Some rocks have fossils.	

Identifying Your Rocks and Minerals

Your specimens are now divided into rocks and minerals. Next you need to perform each of the tests in this list. This will help you decide exactly what types of rocks and minerals your samples are. Remember to write down what you've found in your geology journal.

COLOR

Use a magnifying glass to study each specimen closely. Make a list of any colors you see. Some specimens will have many colors, but others will have only one color.

Feldspar and greenstone

ROCK GUIDE

LUSTER

Luster is the term used to describe how a mineral reflects light. There are many ways to describe the luster of a mineral, and the chart lists the different types. If you believe that a specimen is a mineral, use a magnifying glass to determine its luster. If you believe that a specimen is a rock, skip this step.

HARDNESS

Some rocks and minerals are soft. When you scratch a soft rock with a steel nail, part of the rock can break off or you might see a mark left behind on the rock. Other rocks and minerals are hard. When you scratch a hard rock with a nail, the rock doesn't change. Depending on how hard the rock is,

METALLIC
Metallic minerals look like polished metal. They are very shiny and reflect light very well.

SUBMETALLIC
Submetallic minerals look like metal, but do not look polished. They are not as shiny and do not reflect as much light as metallic specimens.

GLASSY
Glassy minerals look like glass.

PEARLY
Pearly minerals are white and shiny. They look like pearls.

SILKY
Silky minerals look like silk. They have small parallel lines that reflect light.

DULL
Dull minerals are not shiny at all. They reflect very little light, if any at all.

it might even damage the nail. Try scratching each specimen with a nail. If part of the specimen breaks off or a marking is left behind, write "soft" in your journal. If the rock doesn't change, write "hard" in your journal.

STREAK

Streak is the color of a rock or mineral in its powdered form. Interestingly, the streak color is not always the same as the specimen itself. To learn the streak color of a specimen, scrape it once firmly on a streak plate or unglazed tile to make a line. Some specimens will leave a line and others will not. Make sure to examine the tile closely. Some specimens will leave behind the same color as the tile, making the streak color hard to see.

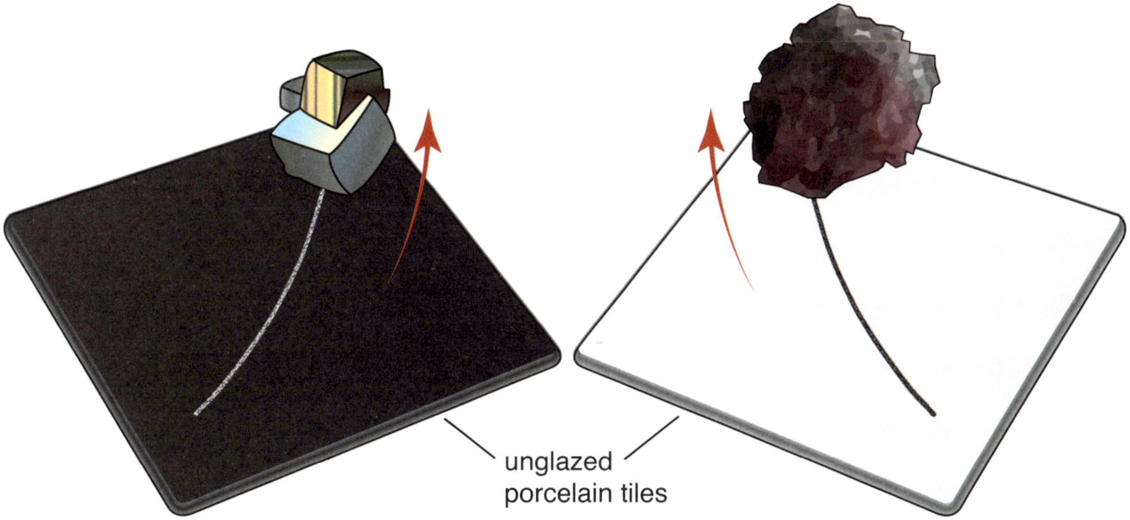

FEATURES

Last, study each specimen closely to see if there are any other characteristics that might help you identify what it is. Do you notice any of the following features in your specimens?

FOSSILS

You've learned that fossils are the remains of plants or animals. Some specimens will contain shells or will have outlines of things that were once living, which are called trace fossils. Remember, fossils are mostly found in sedimentary rocks.

CRYSTALS

Many minerals form crystals. Since rocks are formed from minerals, you will sometimes see crystals in rocks. They may look like specks of glitter throughout the rock.

GRAINS

Some types of rocks are made of sediments that are pressed together. Those sediments are also called grains. Some grains are tiny and hard to see. Other grains are big. Grains can be so big that they look like a bunch of rocks stuck together. The material sticking the grains together is called a matrix or cement.

AIR BUBBLES

Some types of rocks are formed by lava or magma that cools. As the molten rock cools, air bubbles get trapped in the liquid. When the liquid hardens into a rock, those air bubbles appear as tiny holes throughout the rock.

LAYERS

Some rocks form in layers. This creates bands of different colors.

Congratulations! You have studied, tested, and recorded information about the specimens in your collection. Now use this information to figure out the specific types of rocks and minerals in your collection. Use the key that starts on the next page to help you.

IGNEOUS ROCKS

BASALT

COLOR: dark colors (black or gray, sometimes rusty red)

HARDNESS: hard

STREAK: white or gray

WHAT TO LOOK FOR: small crystals

GRANITE

COLOR: light colors (pink, gray, or white)

HARDNESS: hard

STREAK: white

WHAT TO LOOK FOR: large grains and crystals

GABBRO

COLOR: dark colors (black or dark green)

HARDNESS: hard

STREAK: black

WHAT TO LOOK FOR: large grains and crystals

PUMICE

COLOR: light colors (white, beige, or gray)

HARDNESS: hard

STREAK: white, gray, or greenish-white

WHAT TO LOOK FOR: small grains and tiny air bubbles, can float in water

SCORIA

COLOR: dark colors (black, dark gray, or reddish dark brown)

HARDNESS: hard

STREAK: white

WHAT TO LOOK FOR: small grains and large air bubbles

METAMORPHIC ROCKS

GNEISS

COLOR: bands of dark and light colors

HARDNESS: hard

STREAK: white

WHAT TO LOOK FOR: layers and medium grains

MARBLE

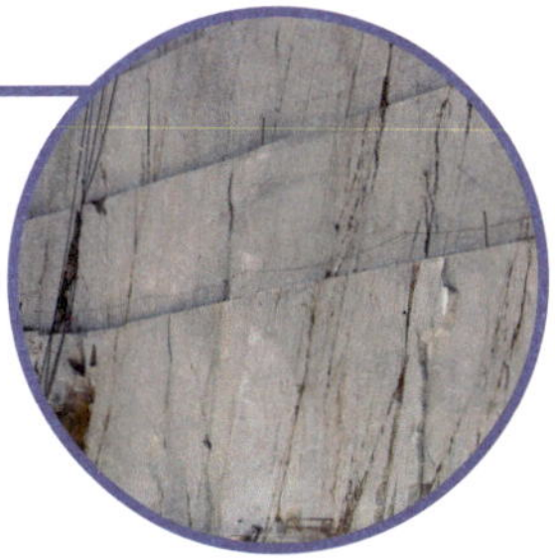

COLOR: light colors (white, gray, green, or pink; rarely black)

HARDNESS: soft

STREAK: white

WHAT TO LOOK FOR: some have tiny crystals, some have swirls or veins of color

ROCK GUIDE

QUARTZITE

COLOR: light colors (white or gray, sometimes pink)

HARDNESS: hard

STREAK: white

WHAT TO LOOK FOR: medium grains and tiny crystals

SCHIST

COLOR: mixed with dark and light colors

HARDNESS: soft

STREAK: white

WHAT TO LOOK FOR: crystals and large grains, the mineral mica, which looks like glitter

SLATE

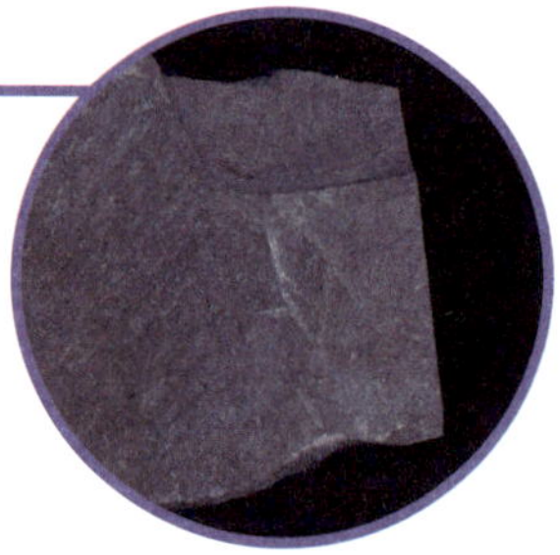

COLOR: light colors (usually light or dark gray, sometimes green, red, black, or purple)

HARDNESS: soft

STREAK: brown

WHAT TO LOOK FOR: tiny grains, can break into thin sheets

SEDIMENTARY ROCKS

COAL

COLOR: dark colors (black or dark brown)

HARDNESS: soft

STREAK: black

WHAT TO LOOK FOR: tiny grains, can burn

LIMESTONE

COLOR: light colors (white, beige, or gray)

HARDNESS: soft

STREAK: white

WHAT TO LOOK FOR: tiny grains and some have fossils

CONGLOMERATE

COLOR: all different colors and shades

HARDNESS: soft

STREAK: white

WHAT TO LOOK FOR: large grains and rocks held together

SANDSTONE

COLOR: dark or light colors (white, red, brown, or black)

HARDNESS: hard

STREAK: white

WHAT TO LOOK FOR: a mixture of tiny and large grains, can have layers

ROCK GUIDE

SHALE

COLOR: dark or light colors (gray, brown, black, or red)

HARDNESS: soft

STREAK: white

WHAT TO LOOK FOR: tiny grains, some have fossils, breaks easily into thin sheets

MINERALS

CALCITE

COLOR: usually white, sometimes colorless, yellow, gray, or brown

LUSTER: glassy

HARDNESS: soft

STREAK: white

WHAT TO LOOK FOR: crystals

FELDSPAR

COLOR: usually white or gray, sometimes colorless, yellow, orange, pink, red, brown, black, blue, or green

LUSTER: glassy

HARDNESS: hard

STREAK: white

WHAT TO LOOK FOR: crystals

GALENA

COLOR: gray or silver

LUSTER: metallic

HARDNESS: soft

STREAK: gray

WHAT TO LOOK FOR: looks like metal

QUARTZ

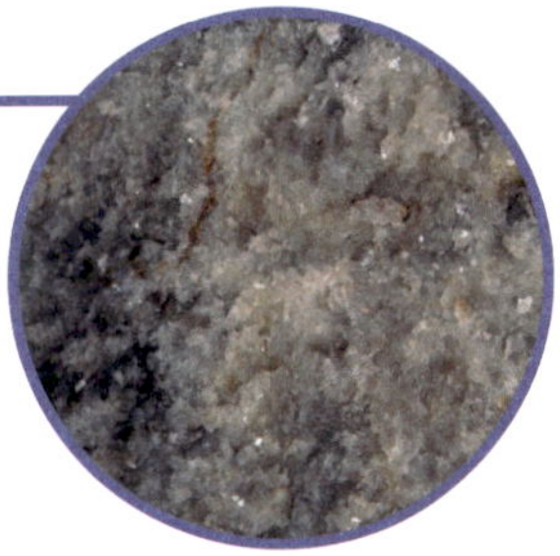

COLOR: usually colorless or white, sometimes purple, blue, pink, orange, or gray

LUSTER: glassy

HARDNESS: hard

STREAK: white

WHAT TO LOOK FOR: crystals

TALC

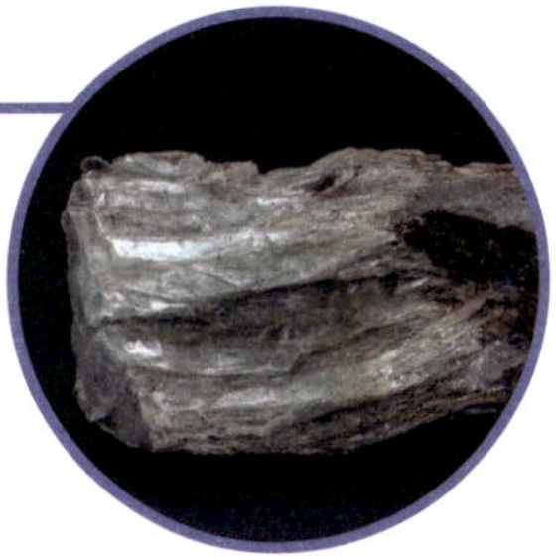

COLOR: white, beige, gray, yellow, brown, or colorless

LUSTER: pearly

HARDNESS: soft

STREAK: white

WHAT TO LOOK FOR: crystals

GROW YOUR OWN CRYSTALS

What You'll Need:

2 CUPS WATER

POT

STOVETOP OR BURNER

TABLE SALT (SEVERAL CUPS)

SPOON

MEDIUM GLASS JAR

SCISSORS

COTTON STRING

PENCIL

What to Do:

1. Ask an adult to boil 2 cups of water in a pot.
2. With the help of an adult, add 1 cup salt to the hot water and stir with a spoon. Once it dissolves, gradually add more salt, a little at a time, until no more salt will dissolve in the water. You will know you are finished when salt crystals stay on the bottom of the pot.
3. Ask an adult to pour the salt water into the jar.
4. Cut a piece of string so that it is about as long as the jar is high. Tie it to the middle of a pencil. Rest the pencil across the top of the jar. The string should hang down into the water, but should not touch the bottom of the jar.
5. Let the string sit in the jar for at least two days. It is important not to move the jar. Carefully remove the string. What do you notice? (If you allow the string to sit for more than two days, the crystals will keep growing larger.)

DID YOU KNOW?

A scientist named Friedrich Mohs came up with a way to figure out how hard a mineral is based on what other minerals can scratch it. The Mohs scale has 10 levels. The higher the number, the harder the mineral. Every mineral on the scale can scratch all the minerals below it. The Mohs scale says that talc is the softest mineral and diamonds are the hardest.

DIG DEEPER!

There are thousands of rocks and minerals. And some are tricky to identify. Are there any in your collection that you could not identify using the key? If so, check out *Smithsonian Handbooks: Rocks and Minerals* by Chris Pellant to find a bigger rock identification key.

Shale formation in Norway

THE ADVENTURE CONTINUES

Even though we know a lot about the Earth, we still have a lot to learn. The Earth is constantly moving and changing. That means there are new discoveries to be made each day. Hopefully this book inspired you to want to learn even more about the planet we call home.

In this book, you learned about Earth's processes and landforms. You learned about its history and its future, how to conduct your own experiments, and how to search for and identify rocks and minerals—but this is just the beginning of your adventure as a junior scientist.

Visit national parks to see Earth's magnificent features up close. Go to natural history museums to examine rocks and fossils. Read the books and visit the websites listed at the end of this book. Observe the world around you in your own backyard or anywhere you travel.

It is never too early to start exploring. By keeping a curious mind and an adventurous spirit, you can be part of the study of geology all your life. A world of discoveries is waiting for you!

GLOSSARY

CANYON: A type of river valley with sides that are almost straight up and flat on top

CONDENSATION: The second stage of the water cycle when water vapor cools, turns back into tiny drops of liquid, and forms clouds

CORE: The central layer of Earth, divided into inner and outer portions

CRUST: The outer layer of Earth, the planet's surface

DESERT: A dry area of land that receives less than 10 inches of rain per year

EARTHQUAKE: Sudden movement in Earth's crust that happens when tectonic plates grind against each other

ELEMENT: A chemical substance that can't be broken down into any other substances

EROSION: The transport of sediments from one place to another

EVAPORATION: The first stage of the water cycle when water changes from a liquid into a gas

FAULT: A large crack in Earth's crust

FOSSIL: The imprint or actual remains of a plant or animal that is captured in rock

GEOLOGY: The study of Earth, the things it is made of, and its history

GLACIER: Large, slow-moving sheet of ice

HILL: Land that is higher than its surroundings, but less than 1,000 feet tall

IGNEOUS ROCK: Rock formed when lava or magma cools and hardens

LANDFORM: The natural shapes of Earth's surface

LAVA: Molten rock that has erupted onto Earth's surface

LITHOSPHERE: The outer, rigid layer of Earth that includes the crust and the upper mantle

MAGMA: Molten rock found below Earth's surface

MANTLE: The middle layer of Earth that contains solid rock hot enough to flow

METAMORPHIC ROCK: Rock that has changed into new rock because of high heat and pressure under the ground

MINERAL: A naturally occurring, inorganic solid with a definite chemical composition and an ordered internal structure

MOUNTAIN: Land that is at least 1,000 feet taller than its surroundings

MOUNTAIN RANGE: A line of connected mountains

PANGAEA: A supercontinent that existed millions of years ago when all the continents were connected

PLAIN: A large, mostly flat area of land

PLATEAU: A tall area of land that is flat on top and has at least one steep, clifflike side

PRECIPITATION: The third stage of the water cycle when water vapor becomes liquid and falls to Earth as rain, snow, sleet, or hail

ROCK CYCLE: The never-ending process of rock erosion, burial, melting, and hardening

SEDIMENT: The small rocks, sand, and minerals that are moved from one place to another by wind, water, or ice

SEDIMENTARY ROCK: Rock formed when sediments are pressed together

SPECIMEN: A sample of something that is used for testing

TECTONIC BOUNDARY: Where two or more tectonic plates meet, either crashing together, moving apart, or sliding along each other

TECTONIC PLATES: The giant moving pieces of Earth's lithosphere

TSUNAMI: A series of waves in the ocean created by underwater volcanoes and earthquakes

VALLEY: A low area of land with steep sides

VOLCANO: An opening in Earth's crust where gas and molten rocks escape from under the surface

WATER CYCLE: The repetitive process of water evaporation, condensation, and precipitation

WATER VAPOR: The gaseous form of water

WEATHERING: When a rock or mineral is broken down into smaller parts by wind, water, ice, or heat

MORE TO DISCOVER

PLACES TO VISIT:

To see one of the world's most dramatic canyons: The Grand Canyon (Arizona)

To see great examples of weathering: Arches National Park (Utah)

To see glaciers and mountains: Glacier Bay National Park (Alaska)

To see volcanoes: Hawai'i Volcanoes National Park (Hawaii)

BOOKS:

All the Water in the World by George Ella Lyon

A Drop Around the World by Barbara Shaw McKinney

National Geographic Kids Everything Rocks and Minerals by Steve Tomecek

Smithsonian Handbooks: Rocks and Minerals by Chris Pellant

The Top of the World: Climbing Mount Everest by Steve Jenkins

WEBSITES:

To see a day-to-day earthquake record: earthquake.usgs.gov

To learn more about geology: kids.nationalgeographic.com/explore/science/geology-101/

To learn more about the U.S. National Parks: nps.gov

INDEX

F

G

H

I

T

U

V

W

Z

ABOUT THE AUTHOR

MEGHAN VESTAL is a former teacher, as well as the founder of the curriculum development company, Vestal's 21st Century Classroom LLC. Meghan has created hands-on curricula for thousands of teachers around the world and designed unique professional development opportunities for new teachers. Her educational programs have been recognized by the United States Congress.